PEARSON   ALWAYS LEARNING

Dennis Sandgathe • Harold Dibble • Paul Goldberg
Shannon McPherron

The Neanderthal Child of Roc de Marsal: A Prehistoric Mystery

Illustrated by: Anna Goldfield

Pearson Learning Solutions, 501 Boylston Street, Suite 900, Boston, MA 02116
A Pearson Education Company
www.pearsoned.com

Printed in Canada

1 2 3 4 5 6 7 8 9 10 XXXX 17 16 15 14

000200010271943614

RD

ISBN 10: 1-323-05635-1
ISBN 13: 978-1-323-05635-6

THE NEANDERTHAL CHILD OF ROC DE MARSAL:

A PREHISTORIC MYSTERY

DENNIS SANDGATHE
HAROLD DIBBLE
PAUL GOLDBERG
SHANNON MCPHERRON

ILLUSTRATED BY:
ANNA GOLDFIELD

HI, EVERYBODY!

MY NAME IS ADAM, AND I'M A PALEO-ANTHROPOLOGIST.

I'M ESPECIALLY INTERESTED IN NEANDERTHALS AND THEIR BEHAVIOR, SO I'M HEADING OUT TO VISIT DIFFERENT NEANDERTHAL OCCUPATION SITES IN EUROPE AND THE MIDDLE EAST TO LEARN AS MUCH AS I CAN!

I'M GOING TO START WITH SITES IN THE DORDOGNE REGION OF SOUTHWEST FRANCE ... JOIN ME!

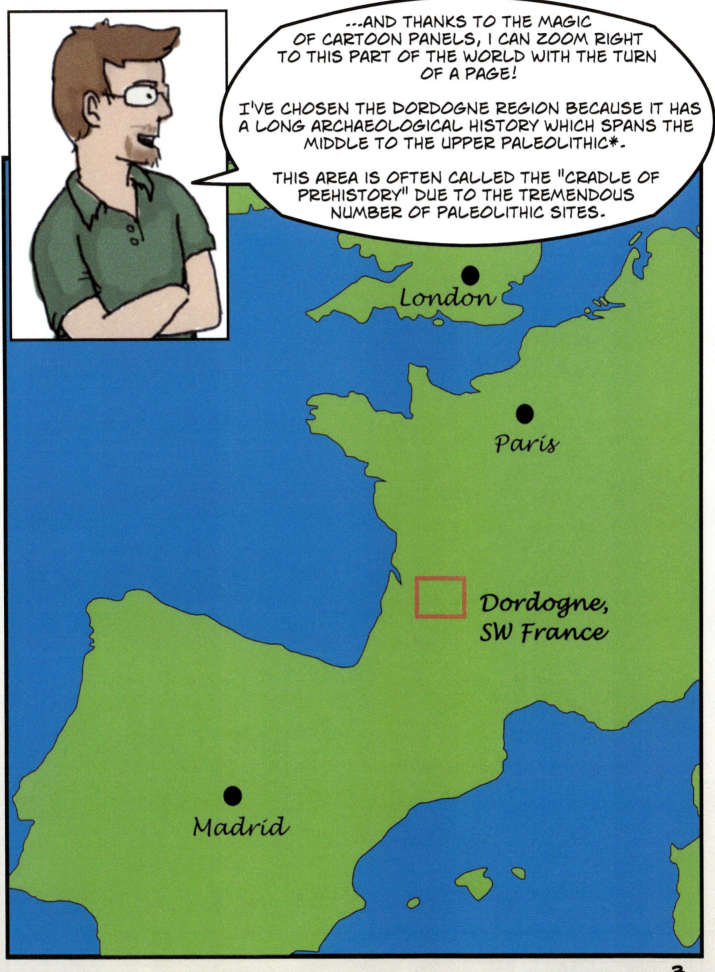

ONE OF THE MOST INTERESTING AND RECENTLY
EXCAVATED OF THESE SITES IS THAT OF ROC DE MARSAL.

THIS IS A SMALL, NATURAL CAVE HIGH IN THE SIDE OF A
LIMESTONE VALLEY.

IT'S QUITE EASY TO ACCESS, IT HAS A NICE, FLAT
TERRACE WHERE PEOPLE COULD LIVE, AND IT PROVIDES
A GREAT VIEW OF THE VALLEY BELOW WHERE PEOPLE
COULD HAVE WATCHED FOR THE MOVEMENT OF GAME.

ROC DE MARSAL WAS FIRST EXCAVATED FROM 1950 TO 1971 BY A LOCAL SCHOOL TEACHER AND AMATEUR ARCHAEOLOGIST, JEAN LAFILLE.

IN 1961, WHILE HE WAS DIGGING HERE, HE DISCOVERED THE SKELETON OF A 2-3 YEAR OLD NEANDERTHAL INFANT - A VERY IMPORTANT DISCOVERY.

HE CALLED IN A VERY FAMOUS FRENCH PREHISTORIAN, FRANÇOIS BORDES FROM THE UNIVERSITY OF BORDEAUX, TO HAVE A LOOK.

BOTH LAFILLE AND BORDES THOUGHT THE BODY OF THE INFANT HAD BEEN INTENTIONALLY BURIED - RATHER THAN HAVING BEEN JUST LEFT THERE WHEN IT DIES AND SUBSEQUENTLY SLOWLY COVERED UP BY NATURALLY ACCUMULATING SEDIMENTS.

THEY THOUGHT THIS MAINLY BECAUSE THE SKELETON WAS RELATIVELY INTACT (ABOUT **80%** OF THE BONES WERE PRESENT) AND BECAUSE LAFILLE FOUND IT IN A DEPRESSION.

AMONG PALEOLITHIC ARCHAEOLOGISTS IT CAME TO BE ONE OF THE MOST WIDELY ACCEPTED EXAMPLES OF NEANDERTHAL INTENTIONAL BURIAL.

AMONG ALMOST ALL LIVING AND RECENT CULTURES THERE ARE WELL DEVELOPED RITUALS FOR DEALING WITH THE REMAINS OF THE DEAD. BURIAL IS ONE OF THE MORE COMMON TYPES OF THESE RITUALS, SO CLEARLY, AT SOME POINT IN THE PAST SOME PEOPLE STARTED BURYING THEIR DEAD.

HOWEVER, THE OLDEST *UNQUESTIONABLE* EXAMPLES OF INTENTIONAL BURIAL DATE ONLY TO WITHIN THE LAST 40,000 YEARS.

SINCE BURIAL OF THE DEAD EVENTUALLY BECAME VERY COMMON, UNDERSTANDING WHEN IT FIRST APPEARED IS OF REAL IMPORTANCE TO UNDERSTANDING THE DEVELOPMENT OF MODERN HUMAN BEHAVIOR ... SO

IMPORTANT QUESTION NUMBER ONE:

WHEN DID BURIAL FIRST APPEAR? AND WHAT DOES IT MEAN?

THE ARCHAEOLOGICAL RECORD IS **NOT** SELF-EVIDENT. THAT IS, WHAT WE THINK WE KNOW ABOUT THE PAST DEPENDS ON HOW ARCHAEOLOGICAL MATERIALS ARE INTERPRETED BY ARCHAEOLOGISTS.

A MAJOR PROBLEM WITH INVESTIGATING THE PREHISTORY OF INTENTIONAL BURIAL IS THAT MOST DATA WERE COLLECTED BEFORE MODERN ARCHAEOLOGICAL METHODS HAD BEEN DEVELOPED. MANY OF THE NEANDERTHAL REMAINS THAT HAVE BEEN CLAIMED TO BE INTENTIONALLY BURIED WERE RECOVERED 100 YEARS AGO OR MORE. AT THAT TIME THERE WERE ALMOST NO TRAINED ARCHAEOLOGISTS. ESSENTIALLY, PEOPLE HAD BEEN DIGGING UP ARCHAEOLOGICAL SITES USING POOR METHODS FOR A LONG TIME, AND MUCH OF THE CURRENT DEBATE STILL RELIES ON THE CLAIMS OF THESE EARLY EXCAVATORS.

WE THINK THAT USING OUR EXCAVATION METHODS TO PROVIDE CONTEXT FOR THE ROC DE MARSAL CHILD WILL GO A LONG WAY TOWARDS ADDRESSING THE DEBATE ON NEANDERTHAL BURIAL.

SO, WHETHER NEANDERTHALS DID BURY THEIR DEAD OR NOT IS AN IMPORTANT QUESTION, BUT THIS ASPECT OF THEIR BEHAVIOR CONTINUES TO BE DEBATED TODAY. AS A PALEOANTHROPOLOGIST I WOULD REALLY LIKE TO KNOW THE TRUTH, IF POSSIBLE, AS IT HAS MAJOR IMPLICATIONS FOR HUMAN PREHISTORY.

A TEAM OF ARCHAEOLOGISTS IS RE-EXCAVATING THE SITE OF ROC DE MARSAL IN ORDER TO TEST THE IDEA OF WHETHER THIS REALLY WAS A BURIAL OR NOT.

THE NEW EXCAVATIONS, USING NEW METHODS AND APPROACHES, COULD POTENTIALLY ANSWER A LOT OF INTERESTING OLD QUESTIONS AND LIKELY GENERATE A BUNCH OF NEW ONES.

LET'S MEET THE TEAM OF RESEARCHERS WHO ARE WORKING HERE AND TRYING TO FIGURE OUT WHAT WAS GOING ON!

HAROLD DIBBLE:
PROFESSOR OF ANTHROPOLOGY AT THE UNIVERSITY OF PENNSYLVANIA SPECIALIZING IN THE ANALYSIS OF STONE TOOLS.

SHANNON MCPHERRON:
AMERICAN RESEARCHER AT MAX PLANCK INSTITUTE IN LEIPZIG, GERMANY ALSO SPECIALIZING IN STONE TOOL ANALYSIS.

ALAIN TURQ:
CHIEF CURATOR OF HERITAGE AT THE NATIONAL MUSEUM OF PREHISTORY OF FRANCE. AN ARCHAEOLOGIST AND GEOLOGIST SPECIALIZING IN IDENTIFYING FLINT SOURCES.

PAUL GOLDBERG:
PROFESSOR OF ARCHEOLOGY AT BOSTON UNIVERSITY. A GEOLOGIST SPECIALIZING IN THE MICROSCOPIC ANALYSIS OF SEDIMENTS.

DENNIS SANDGATHE:
LECTURER IN ARCHAEOLOGY AT
SIMON FRASER UNIVERSITY, CANADA.
HE SPECIALIZES IN THE ANALYSIS OF
STONE TOOLS.

TERESA STEELE:
PROFESSOR OF ANTHROPOLOGY,
UNIVERSITY OF CALIFORNIA, DAVIS. A
ZOOARCHAEOLOGIST: SHE ANALYZES
ANIMAL BONES FROM ARCHEOLOGY
SITES.

VERA ALDEIAS:
PORTUGUESE GEOARCHAEOLOGIST
WORKING AT THE MAX PLANCK
INSTITUTE IN LEIPZIG, GERMANY.

VIRGINIE SINET-MATHIOT:
GRADUATE STUDENT AT THE UNIVERSITY
OF BORDEAUX. SHE MANAGES THE LAB
WHERE THE ARTIFACTS FROM THE SITE
ARE ANALYZED.

OUR CREW IS MADE UP OF STUDENTS AND SPECIALISTS FROM ALL OVER THE WORLD. THIS MAP INDICATES THE MANY COUNTRIES AND INSTITUTIONS THAT OUR CREW MEMBERS HAVE COME FROM OVER THE YEARS.

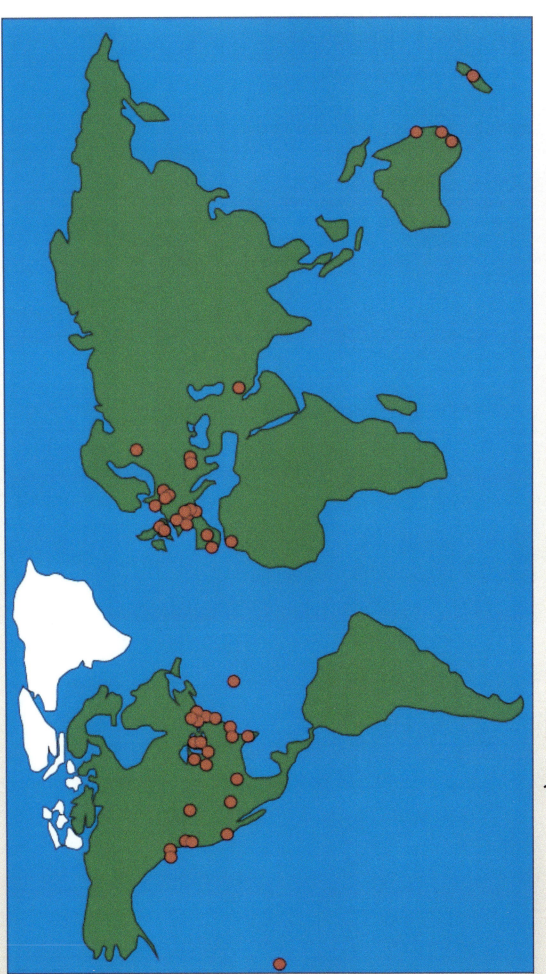

IT'S GREAT TO HAVE SO MANY DIFFERENT BACKGROUNDS AND PERSPECTIVES ON OUR TEAM.

PREHISTORIANS HAVE BEEN ARGUING FOR YEARS ABOUT WHETHER OR NOT NEANDERTHAL CULTURE COULD BE DEFINED AS "MODERN".

MODERN CULTURE IS TYPICALLY REFLECTED IN LANGUAGE, ART, AND RELIGION.

OF COURSE, SINCE NEANDERTHALS DISAPPEARED ABOUT 30,000 YEARS AGO, THERE ARE NONE AROUND TODAY THAT WE CAN STUDY. WE CAN ONLY EXAMINE THE MATERIALS – BONES, LITHIC ARTIFACTS, HEARTHS, AND SEDIMENTS – THAT THEY LEFT BEHIND.

THE INTERPRETATION OF THESE MATERIALS HAVE CONTRIBUTED A GREAT DEAL TO OUR UNDERSTANDING OF HOW NEANDERTHALS LIVED, DAY-TO-DAY AND OVER DECADES THROUGH THE MILLENNIA WHEN THEY WERE AROUND, BUT IT IS STILL INCREDIBLY DIFFICULT TO DETERMINE HOW THEY THOUGHT AND COMMUNICATED.

SOME PEOPLE ARGUE THAT NEANDERTHALS WERE NOT VERY DIFFERENT FROM US MODERN HUMANS IN THEIR CULTURE AND BEHAVIOR: THAT THEY WERE GENERALLY MUCH LIKE THE HUNTER-GATHERER PEOPLE WE CAN STILL SEE TODAY IN PARTS OF AFRICA, NEW GUINEA, AUSTRALIA, AND SOUTH AMERICA.

AT THE OTHER END OF THE SPECTRUM OF OPINION ARE THOSE WHO SAY THAT NEANDERTHALS WERE INCAPABLE OF HUMAN-LIKE BEHAVIOR, BUT RATHER WERE MORE LIKE OTHER ANIMALS THAT LIVE IN GROUPS--LIKE WOLVES, OR CHIMPANZEES.

CHIMPANZEES USE SIMPLE TOOLS TO OBTAIN FOOD, AND WOLVES HAVE A COMPLEX "PACK" SOCIAL STRUCTURE AND HUNT COHESIVELY AS A GROUP. BOTH TOOL USE AND SOCIAL COMPLEXITY ARE ASPECTS THAT WE ASSOCIATE WITH "MODERNITY," AND YET WE CERTAINLY WOULD NOT THINK OF A GROUP OF WOLVES OR CHIMPANZEES AS HAVING THE SAME COGNITIVE CAPABILITIES AS MODERN HUMANS. WHY IS THIS? WHAT MAKES MODERN HUMANS *DIFFERENT*?

How close to modern humans were Neanderthals in their culture and behavior??

What's REALLY going on in there?

One way to address this question is to look for evidence of abstract thinking in Neanderthal sites.

WHERE DO WE COME FROM?

HOW DO I COMMUNICATE MY IDEAS TO OTHERS?

WHAT HAPPENS WHEN WE DIE?

ONE OF THE BEST POTENTIAL INDICATORS OF ABSTRACT THINKING AND HUMAN-LIKE BEHAVIOR IS THE DELIBERATE BURIAL OF THE DEAD.

AT EARLY MODERN HUMAN SITES IN ISRAEL THERE ARE CLAIMS FOR INTENTIONAL BURIALS DATED TO AS EARLY AS 100,000 YEARS AGO!

THESE BODIES APPEAR TO HAVE BEEN BURIED WITH SIMPLE GRAVE GOODS LIKE SHELL BEADS AND RED OCHRE.

IF THIS IS TRUE, IT QUITE POSSIBLY REFLECTS SOME SORT OF BELIEF SYSTEM – 'RELIGION' OF SOME SORT – WHICH INDICATES MODERN CULTURE.

SO, IF IT CAN BE SHOWN THAT AT NEANDERTHAL SITES THERE WERE BURIALS, THEN PERHAPS THERE WAS RELIGION AS WELL.

THE PROPER SCIENTIFIC APPROACH

HOWEVER, FOR MANY OF THE EXAMPLES OF NEANDERTHAL SKELETONS DISCOVERED IN PALAEOLITHIC SITES, EVEN WHEN THERE IS NO DIRECT EVIDENCE THAT IT WAS AN INTENTIONAL BURIAL, SOME RESEARCHERS HAVE MADE THE ARGUMENT:

"WHO KNOWS WHAT AN INTENTIONAL NEANDERTHAL BURIAL MIGHT LOOK LIKE? THIS COULD HAVE BEEN ONE AND SO IT PROBABLY IS".

THIS IS NOT A LEGITIMATE SCIENTIFIC ARGUMENT. WOULD IT BE REASONABLE TO ARGUE THAT NEANDERTHALS PRACTICED CREMATION JUST BECAUSE WE FOUND A POCKET OF ASH IN THE DEPOSITS OF A SITE? THE ASH **COULD** BE THE REMAINS OF A CREMATED NEANDERTHAL. HOWEVER, IT **COULD** ALSO BE LOTS OF OTHER THINGS; BURNED ANIMAL BONE OR BURNED WOOD. WE COULD EVEN SAY THAT NEANDERTHALS MIGHT HAVE PLAYED THE PIANO, BUT SHOULDN'T WE WAIT UNTIL WE FIND ACTUAL PIANOS BURIED IN THEIR SITES BEFORE WE ACCEPT THIS AS TRUE!

THE BOTTOM LINE IS THAT THERE ARE A NUMBER OF DIFFERENT WAYS THAT THE REMAINS OF A NEANDERTHAL COULD END UP BURIED IN A SITE AND INTENTIONAL BURIAL IS JUST ONE OF THEM. LIKE ALL SCIENTIFIC HYPOTHESES, THE SUGGESTION THAT THE REMAINS OF A NEANDERTHAL WERE INTENTIONALLY BURIED NEEDS TO BE TESTED AND NEEDS POSITIVE SUPPORTING EVIDENCE FOR IT TO BE ACCEPTED. WITHOUT ACTUAL EVIDENCE A HYPOTHESIS REMAINS SIMPLE CONJECTURE AND WE CANNOT BASE SUBSEQUENT INTERPRETATIONS ON CONJECTURE: IF WE DID THAT SCIENCE WOULD BE A HUGE HOUSE OF CARDS READY TO COLLAPSE AT ANY TIME.

ONE OF THE MORE CONVINCING EXAMPLES OF POTENTIAL NEANDERTHAL BURIAL IS AT KEBARA CAVE ON MT CARMEL IN ISRAEL.

A GEOARCHAEOLOGICAL STUDY OF THE CONTEXT OF THE KEBARA NEANDERTHAL SKELETON, DATED TO ABOUT 60,000 YEARS AGO, SUGGESTED THAT THE INDIVIDUAL WAS LAID IN A PIT SPECIALLY DUG FOR ITS INTERMENT.

SEDIMENTS OUTSIDE THE PIT DIFFERED IN COLOUR AND COMPOSITION FROM THOSE THAT COVERED THE SKELETON AND FILLED THE PIT. AS THE PIT WAS A CLOSED DEPRESSION AND NOT A CHANNEL, THIS OBSERVATION MEANS THE PIT WAS PROBABLY NOT FILLED BY NATURAL DEPOSITION – IT HAD LIKELY BEEN FILLED RAPIDLY.

ALSO, THE SEDIMENTS THAT FILLED THE PIT WERE NOT LAYERED, AS NATURALLY ACCUMULATED SEDIMENTS ARE. RATHER, THE PIT WAS FILLED WITH A SINGLE, HOMOGENEOUS SEDIMENT.

THIS IS WHAT WE WOULD EXPECT IF THE BODY WAS INTENTIONALLY PLACED IN THE PIT AND THEN COVERED UP IMMEDIATELY.

WE WILL TALK MORE ABOUT THIS LATER.

IF THE SKELETON AT KEBARA SHOWS SIGNS OF DELIBERATE BURIAL, BUT OTHER NEANDERTHAL SKELETONS DO NOT, HOW DO WE GO ABOUT CLASSIFYING THE WAY DIFFERENT GROUPS OF NEANDERTHALS DEALT WITH THEIR DEAD?

IF NOT ALL NEANDERTHALS BURIED THEIR DEAD OR IF INTENTIONAL BURIAL WAS NOT A COMMON BEHAVIOR AMONG NEANDERTHALS, WHAT DOES THIS TELL US ABOUT NEANDERTHAL CULTURE?

THIS IS WHY A REEXAMINATION OF THE CONTEXT OF THE ROC DE MARSAL INFANT, ALONG WITH ALL OTHER CLAIMS FOR NEANDERTHAL BURIAL, IS SO IMPORTANT.

THE IMPLICATIONS OF OTHER SITES WITH DELIBERATE BURIAL MIGHT TELL US MORE ABOUT RELIGION, CULTURE, AND THE DIFFERENCES BETWEEN MODERN HUMANS AND NEANDERTHALS.

IT IS POSSIBLE THAT NEANDERTHALS SOMETIMES BURIED THEIR DEAD SIMPLY BECAUSE THEY DID NOT WANT THE REMAINS OF THEIR FAMILY MEMBERS TO BE EATEN BY SCAVENGERS. THAT IS TO SAY THAT INTENTIONAL BURIAL DOES NOT AUTOMATICALLY MEAN THEY HAD SOME SORT OF SUPERNATURAL BELIEFS.

HOWEVER, IF NEANDERTHALS DID HAVE SOME SORT OF RELIGION, THEN, BY DEFINITION, IT HAD TO BE A SYMBOLIC SYSTEM SHARED AMONG DIFFERENT PEOPLE. THIS IS THE WAY CULTURE IS SHARED BY MODERN HUMANS TODAY: THROUGH THE USE OF ABSTRACT SYMBOLS AND IDEAS.

SO BY LOOKING FOR EVIDENCE OF NEANDERTHAL BURIALS, WE'RE LOOKING FOR ANY EVIDENCE OF ABSTRACT THINKING, WHICH WOULD INDICATE A MAJOR SIMILARITY BETWEEN THEM AND US ...

... AND, IN ORDER TO MAKE SURE THAT WE UNDERSTAND THE NATURE OF A BURIAL AS COMPLETELY AS POSSIBLE, WE NEED TO EXAMINE THE ENTIRE CONTEXT OF ANY SKELETON. THIS INCLUDES THE NATURE OF THE DEPOSITS IT WAS FOUND IN AND ANY MATERIAL FINDS ASSOCIATED WITH IT.

WE NEED ACTUAL, *POSITIVE EVIDENCE* OF HOW IT GOT THERE.

SO, LET'S GO BACK TO ROC DE MARSAL AND SEE WHAT THE TEAM WORKING THERE CAN TELL US ABOUT THE QUESTION OF NEANDERTHAL INTENTIONAL BURIAL.

THE ROC DE MARSAL CAVE IS ACTUALLY PART OF A LARGER CAVE SYSTEM THAT WAS FORMED BY DISSOLUTION OF THE LIMESTONE ALONG A NATURAL JOINT IN THE BEDROCK, WHICH RUNS PERPENDICULAR TO THE CLIFF FACE SEEN HERE.

WESTERN EUROPE HAS MANY THOUSANDS OF CAVES LIKE THIS.

WE CAN TELL FROM LAFILLE'S INITIAL PUBLICATION THAT HE AND BORDES DEFINITELY INTERPRETED THE FIND AS A DELIBERATE BURIAL BECAUSE OF HOW COMPLETE THE SKELETON WAS. HOWEVER, WHAT IS STILL NOT CLEAR IS IF THEY THOUGHT THE PIT HAD BEEN DELIBERATELY CREATED FOR THE BURIAL.

LATER PUBLICATIONS BY OTHER RESEARCHERS (MOST OF WHOM DIDN'T ACTUALLY WORK AT THE SITE) CLAIMED THE BODY HAD DEFINITELY BEEN IN AN INTENTIONALLY DUG PIT.

UNFORTUNATELY, THE BLOCK OF SEDIMENT SURROUNDING THE SKELETON WAS COMPLETELY REMOVED BY LAFILLE, SO WE'LL NEED TO EXAMINE THE INTACT PARTS OF THE SITE TO UNDERSTAND THE NATURE OF THE BEDROCK FLOOR AND THE SEDIMENTS THAT ACCUMULATED ON IT. THIS COULD TELL US WHETHER THE PIT WAS NATURAL OR ANTHROPOGENIC.

IF THE PIT WERE A NATURAL FEATURE IT MAKES THE ARGUMENT FOR INTENTIONAL BURIAL WEAKER, BUT WE STILL NEED TO UNDERSTAND THE STRATIGRAPHIC CONTEXT OF THE SKELETON IN ORDER TO BETTER UNDERSTAND HOW IT GOT THERE. THIS WILL BE ONE OF OUR MAJOR GOALS.

23

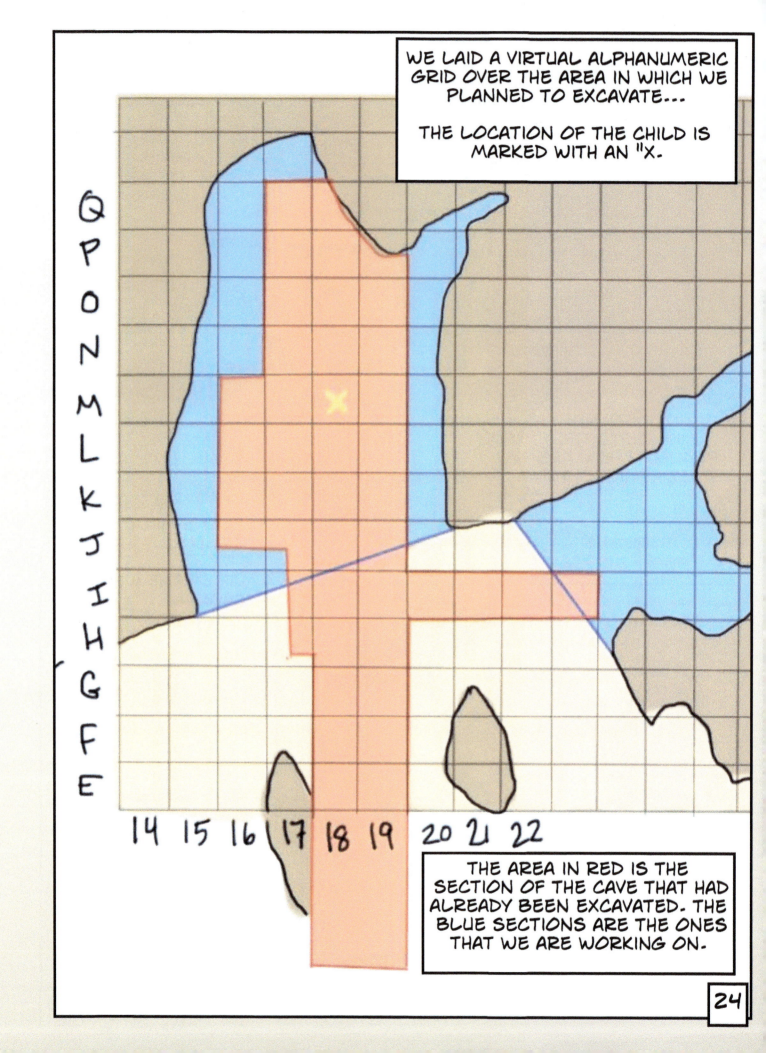

HERE'S WHAT THE CAVE FLOOR AND ARCHAEOLOGICAL SEDIMENTS LOOK LIKE AROUND THE SPOT WHERE THE SKELETON WAS EXCAVATED (MARKED WITH AN "X").

WE'LL SEE THIS PHOTO AGAIN LATER WHEN WE DISCUSS THE ISSUE OF THE "BURIAL."

HMM, INTERESTING! BUT I STILL HAVE SOME QUESTIONS ABOUT HOW YOU EXCAVATE THE SITE AND THE SORTS OF THINGS YOU FIND IN THE SEDIMENTS.

26

THE TOTAL STATION
...AN ARCHAEOLOGIST'S BEST FRIEND

THE TOTAL STATION IS A COMBINATION OF A DIGITAL THEODOLITE, WHICH MEASURES SLOPES AND ANGLES, AND AN ELECTRONIC DISTANCE METER (EDM).

THIS TOOL USES A LASER BEAM TO RECORD DATA POINTS IN THEIR PRECISE SPATIAL LOCATIONS.

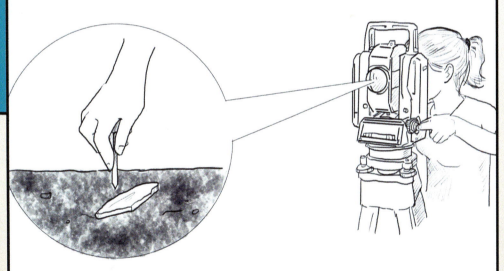

THE TOTAL STATION MEASURES EVERYTHING IN REFERENCE TO A PERMANENT POINT THAT IS ESSENTIALLY UNIQUE TO EACH SITE, KNOWN AS A 'DATUM'.

THE TOTAL STATION SOFTWARE STORES THE COORDINATES OF THE DATUM, AND CALCULATES ALL OTHER POINT COORDINATES IN REFERENCE TO THE DATUM.

HERE SOMEONE'S HAND IS DIRECTING THE TOTAL STATION OPERATOR TO RECORD THE 3D COORDINATES OF AN ARTIFACT THAT WAS JUST EXPOSED.

HERE'S HOW THE TOTAL STATION WORKS:

THE INSTRUMENT SENDS OUT A LASER BEAM, WHICH BOUNCES OFF THE OBJECT BEING MEASURED.

THE TOTAL STATION THEN MEASURES THE ANGLE OF THE RETURNING BEAM, ALONG WITH ITS WAVELENGTH AND FREQUENCY, AND TRANSLATES THAT INFORMATION INTO DISTANCE AND SLOPE.

THIS TELLS THE TOTAL STATION THE EXACT LOCATION OF THAT OBJECT IN 3 DIMENSIONAL SPACE.

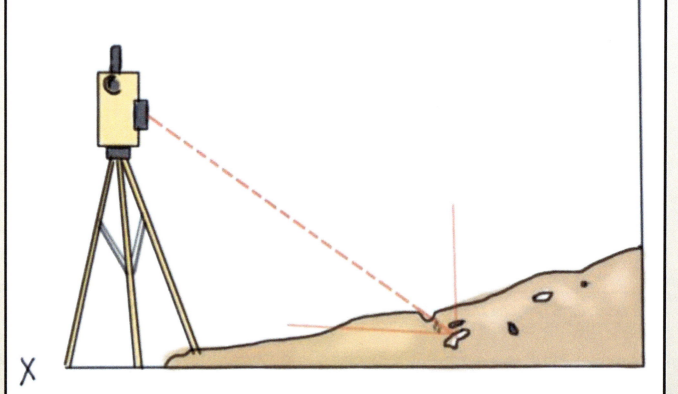

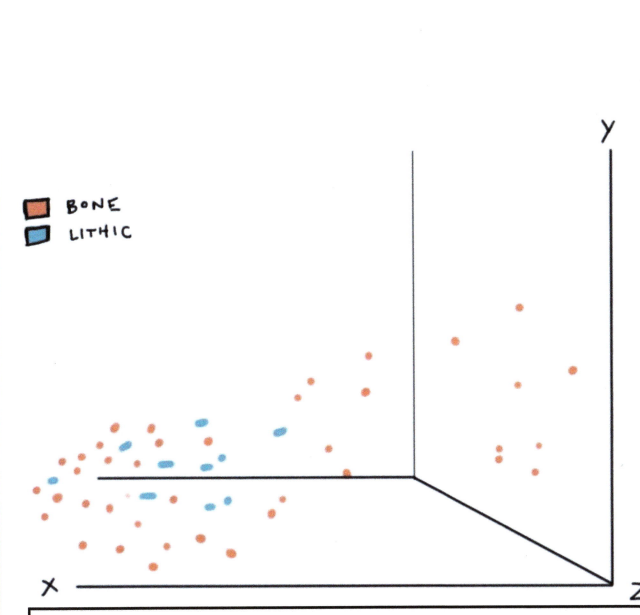

DATA POINTS – WHICH CAN BE STONE ARTIFACTS, BONES, SAMPLES, OR TOPOGRAPHIC ELEVATIONS – ARE PLOTTED DIGITALLY INTO A CARTESIAN GRID. EACH INDIVIDUAL POINT HAS X, Y, AND Z COORDINATES.

ONCE ALL THE DATA HAVE BEEN PLOTTED, USING A APPROPRIATE COMPUTER SOFTWARE, WE CAN PRODUCE A MAP OF ALL THE POINTS.

IT'S ALMOST LIKE HAVING X-RAY VISION! YOU CAN SEE WHERE ALL THE PLOTTED OBJECTS LAY WITHIN THE SITE JUST AS THEY WERE BEFORE BEING EXCAVATED.

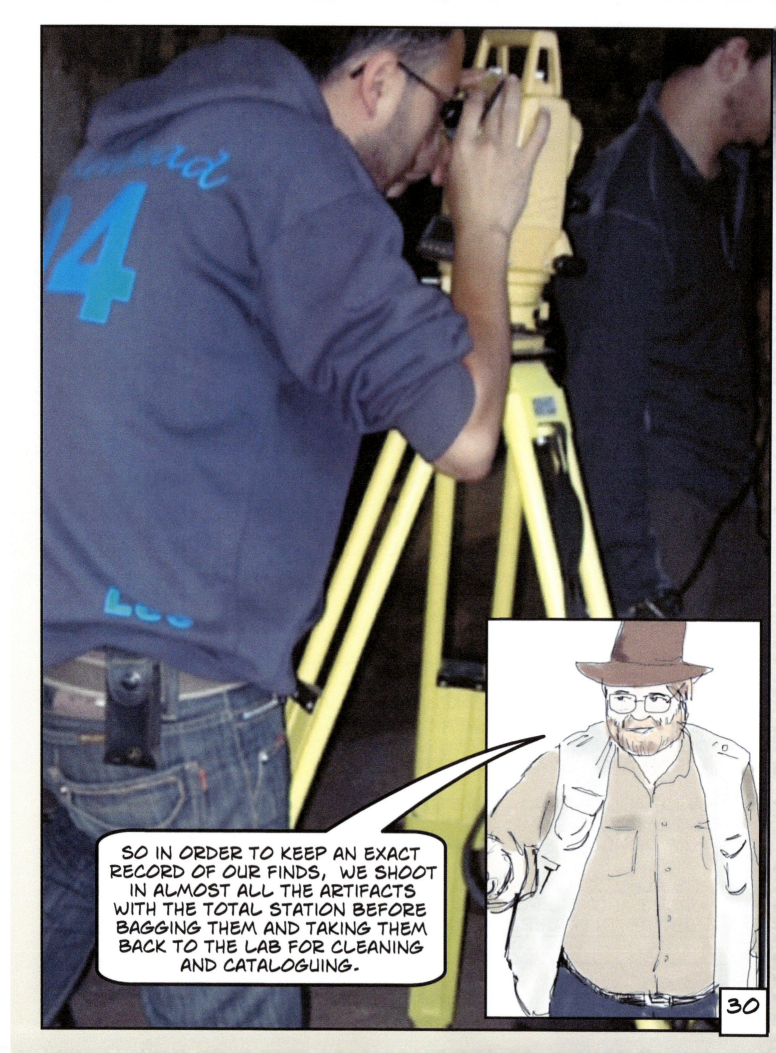

SO IN ORDER TO KEEP AN EXACT RECORD OF OUR FINDS, WE SHOOT IN ALMOST ALL THE ARTIFACTS WITH THE TOTAL STATION BEFORE BAGGING THEM AND TAKING THEM BACK TO THE LAB FOR CLEANING AND CATALOGUING.

FLINT IS A TYPE OF SEDIMENTARY ROCK, COMPOSED MAINLY OF SILICA, THAT FORMS IN DEEP OCEAN DEPOSITS. THESE DEPOSITS EVENTUALLY TURN INTO LIMESTONE IN WHICH NODULES OF FLINT ARE FOUND. LIMESTONE IS AN ABUNDANT COMPONENT OF THE GEOLOGY OF WESTERN EUROPE. FLINT IS VERY COMMON HERE AND IT IS A GREAT RAW MATERIAL FOR MAKING STONE TOOLS WITH SHARP, DURABLE EDGES.

NEANDERTHALS COULD EITHER PICK UP NODULES FROM THE GROUND THAT HAD NATURALLY ERODED OUT OF LIMESTONE OR THEY COULD GET THEM FROM RIVER BEDS.

FLINT FROM DIFFERENT FORMATIONS OFTEN HAVE SLIGHTLY DIFFERENT CHARACTERISTICS. THESE CHARACTERISTICS INCLUDE COLOUR, THE WAY IN WHICH IT WEATHERS, AND THE TYPES OF MICROFOSSILS IT CONTAINS.

ARCHAEOLOGISTS CAN OFTEN DETERMINE WHERE THE FLINT FROM A SPECIFIC SITE OR LAYER CAME FROM BASED ON ITS CHARACTERISTICS.

THIS CAN BE VERY USEFUL INFORMATION BECAUSE KNOWING HOW FAR FLINT WAS BEING MOVED BY NEANDERTHALS CAN HELP US UNDERSTAND HOW MUCH THEY WERE MOVING AROUND THE LANDSCAPE AS THEY MOVED FROM SITE TO SITE HUNTING FOR GAME.

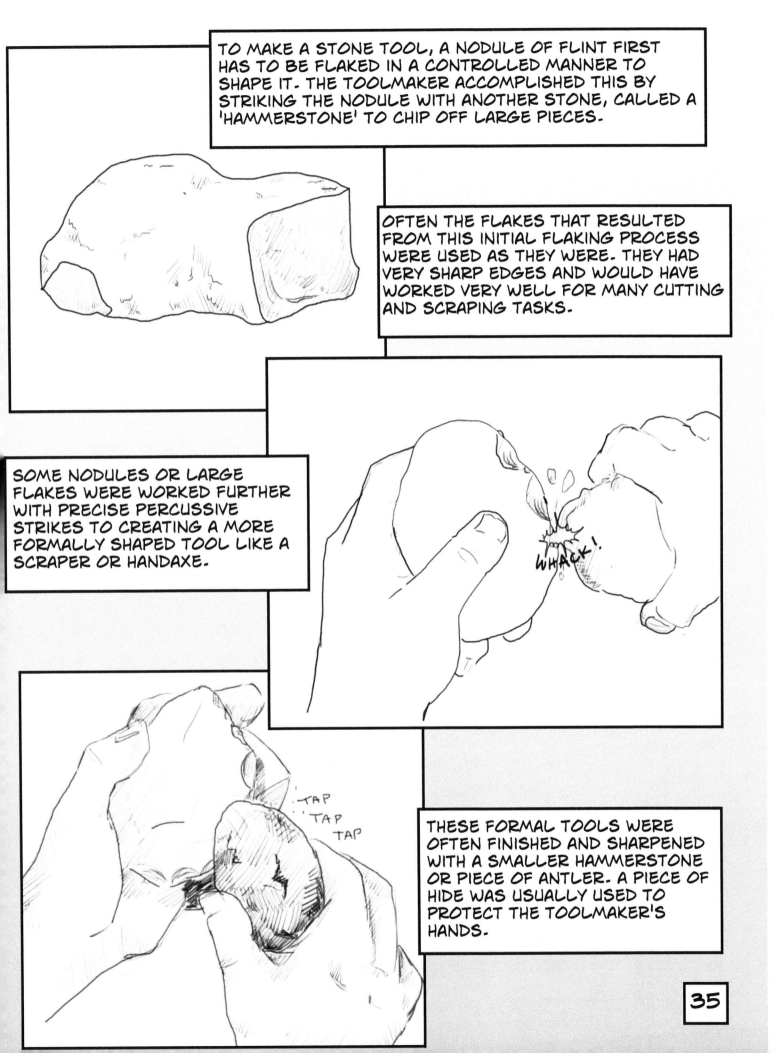

TO MAKE A STONE TOOL, A NODULE OF FLINT FIRST HAS TO BE FLAKED IN A CONTROLLED MANNER TO SHAPE IT. THE TOOLMAKER ACCOMPLISHED THIS BY STRIKING THE NODULE WITH ANOTHER STONE, CALLED A 'HAMMERSTONE' TO CHIP OFF LARGE PIECES.

OFTEN THE FLAKES THAT RESULTED FROM THIS INITIAL FLAKING PROCESS WERE USED AS THEY WERE. THEY HAD VERY SHARP EDGES AND WOULD HAVE WORKED VERY WELL FOR MANY CUTTING AND SCRAPING TASKS.

SOME NODULES OR LARGE FLAKES WERE WORKED FURTHER WITH PRECISE PERCUSSIVE STRIKES TO CREATING A MORE FORMALLY SHAPED TOOL LIKE A SCRAPER OR HANDAXE.

WHACK!

TAP
TAP
TAP

THESE FORMAL TOOLS WERE OFTEN FINISHED AND SHARPENED WITH A SMALLER HAMMERSTONE OR PIECE OF ANTLER. A PIECE OF HIDE WAS USUALLY USED TO PROTECT THE TOOLMAKER'S HANDS.

35

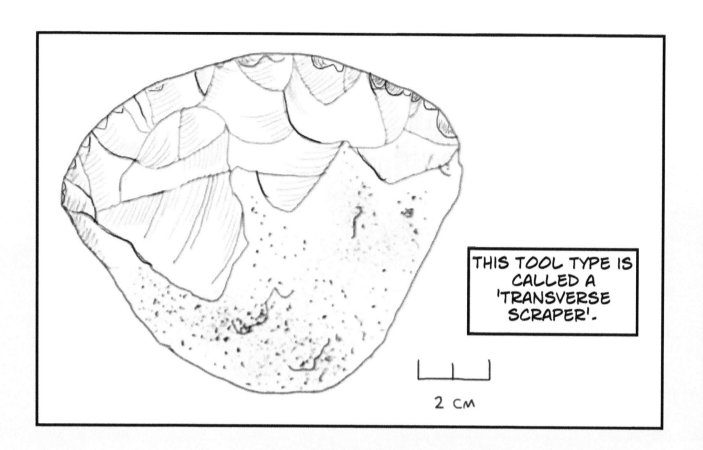

THIS TOOL TYPE IS CALLED A 'TRANSVERSE SCRAPER'.

2 CM

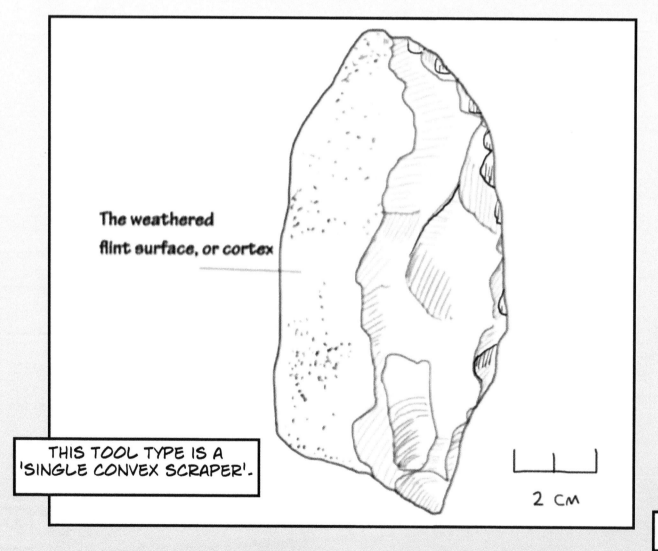

The weathered flint surface, or cortex

THIS TOOL TYPE IS A 'SINGLE CONVEX SCRAPER'.

2 CM

NEANDERTHALS ALSO SOMETIMES MADE SMALL HANDAXES LIKE THIS.

THEY WERE LIKELY USED FOR A VARIETY OF TASKS, INCLUDING BUTCHERING ANIMALS.

STONE TOOLS WERE HARD ENOUGH TO CHOP THROUGH BONE, BUT THEY WERE ALSO SHARP ENOUGH TO SLICE THROUGH HIDE.

THIS MADE FOR A VERSATILE IMPLEMENT THAT WAS ALSO PORTABLE.

IT'S COOL THAT THESE TOOLS COULD BE SO VERSATILE, ESPECIALLY FOR HUNTING AND BUTCHERING!

THAT BRINGS ME TO MY NEXT QUESTION:

I'M CURIOUS ABOUT THE ANIMALS THAT WERE ACTUALLY HUNTED HERE, AND THE ENVIRONMENT THEY INHABITED.

FOR MORE ON THAT, WE'LL TURN TO OUR FAUNA EXPERT, TERESA STEELE. SHE STUDIES ZOOARCHAEOLOGY, WHICH CONTRIBUTES A GREAT DEAL TO OUR UNDERSTANDING OF PALEOLITHIC DIETS.

39

CLIMATE FLUCTUATED A LOT OVER THE PAST 125,00 YEARS:
A PERIOD GEOLOGISTS CALL THE LATE PLEISTOCENE.

FOR THE FIRST 50,000 YEARS OF THE LATE PLEISTOCENE
THE EARTH WAS IN AN INTERGLACIAL: THE CLIMATE WAS
MAINLY WARM AND WET (TODAY WE ARE IN ANOTHER
INTERGLACIAL): ALTHOUGH THERE WERE STILL SOME COLD
PERIODS. DURING INTERGLACIALS MUCH OF WESTERN
EUROPE WAS COVERED IN FORESTS OF MAINLY DECIDUOUS
TREE SPECIES.

HOWEVER, STARTING AROUND 75,000 YEARS AGO THE
EARTH'S TEMPERATURES BEGAN TO RAPIDLY DROP AND
GLACIERS ADVANCED ACROSS NORTHERN EUROPE AND IN THE
MOUNTAINS. THE CLIMATE BECAME COLD AND DRY AND MUCH
OF WESTERN EUROPE LOOKED LIKE ARCTIC TUNDRA.

DURING WARMER, WETTER PERIODS WESTERN EUROPE WAS INHABITED BY FOREST-ADAPTED SPECIES LIKE RED DEER, ROE DEER, AND WILD PIGS.

DURING GLACIAL PERIODS, THE FOREST SPECIES WERE REPLACED BY SPECIES WELL ADAPTED TO COLDER CONDITIONS AND OPEN GRASSLAND: MAMMOTH, HORSE, BISON, AND REINDEER.

WHEN WE IDENTIFY SPECIES FOUND IN ARCHAEOLOGICAL LAYERS, WE CAN INFER WHAT THE LOCAL ENVIRONMENT WOULD HAVE BEEN LIKE AT THE TIME THAT THESE ANIMALS WERE HUNTED.

Percentage of each species in each level

Level	Age (years ago)	Red Deer	Roe Deer	Wild Pig	Reindeer	Horse	Bison
2	50,000						
3							
4	75,000						
5							
6							
7							
8							
9	85,000						

COLDER ←→ WARMER

SO IN IDENTIFYING ANIMAL BONES IN A PARTICULAR ARCHAEOLOGICAL LEVEL, WE'RE LOOKING FOR WHICH ANIMALS ARE THE MOST COMMON. THAT CAN GIVE US SOME IDEA OF THE PREVAILING ENVIRONMENT ASSOCIATED WITH THAT LEVEL AND WHAT SORT OF CLIMATE PEOPLE HAD TO DEAL WITH AT THE TIME IT WAS BEING DEPOSITED.

HERE, TWO CREW MEMBERS ARE EXCAVATING A PART OF LAYER 4.

THE FAUNA IN THIS LAYER IS COMPOSED ALMOST ENTIRELY OF THE REMAINS OF REINDEER, INDICATING THAT THIS LAYER WAS DEPOSITED DURING A COLD, DRY PERIOD.

THESE REINDEER BONES HAVE BEEN VERY HEAVILY BUTCHERED. NEANDERTALS BROKE THE BONES UP INTO VERY SMALL PIECES, PROBABLY TO GET HIGHLY NUTRITIOUS MARROW OUT OF THEM.

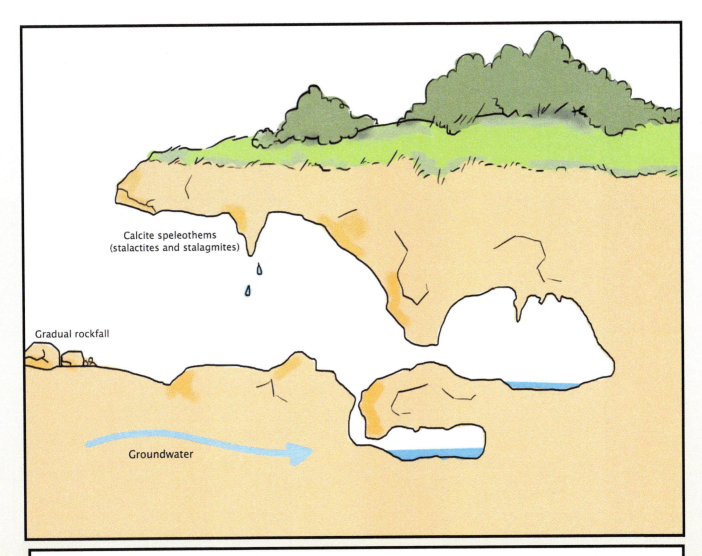

Calcite speleothems
(stalactites and stalagmites)

Gradual rockfall

Groundwater

KARSTIC CAVES ARE FORMED BY THE DISSOLUTION OF LIMESTONE BEDROCK, BY FIRST GROUNDWATER AND LATER FROM WATER PERCOLATING FROM THE SURFACE. THESE PROCESSES RESULT IN IRREGULAR CAVITIES, CHANNELS, AND CHAMBERS, DEPENDING ON FACTORS SUCH AS THE TYPE OF BEDROCK, HOW IT IS FRACTURED, CLIMATE, AND GROUNDWATER TABLE.

KARSTIC CAVE AND TUNNEL SYSTEMS CAN EXTEND FOR TENS OF KILOMETERS UNDER THE EARTH'S SURFACE; THE CAVE AT ROC DE MARSAL IS ACTUALLY PART OF A MUCH LARGER LOCAL SYSTEM, ALTHOUGH IT ITSELF FORMED FROM DISSOLUTION ALONG A JOINT IN THE BEDROCK.

CAVES TRAP SEDIMENTS FROM HUMAN ACTIVITIES, AS WELL AS THOSE FROM NATURAL PROCESSES SUCH AS FLOODING, DRIPPING WATER, OR WATER WASHING SEDIMENTS INTO THE INTERIOR. THE PROTECTED NATURE OF CAVES AND ROCK SHELTERS MAKES THEM IDEAL PLACES FOR BOTH HUMANS AND ANIMALS TO FIND SHELTER. THIS FEATURE AND THE PROTECTION OF THE SEDIMENTS PROVIDED BY THE SHELTER OF THE CAVE, MEANS THAT PREHISTORIC CAVE SITES ARE MUCH MORE LIKELY TO SURVIVE AND BE VISIBLE THAN OPEN-AIR SITES.

HERE IS AN ENLARGED FISSURE LEADING TOWARD AN ADJACENT CAVE. IT IS FILLED WITH SEDIMENTS WASHED AND BLOWN IN FROM OUTSIDE AND CONTAINS THE REMAINS OF SEVERAL HYAENA COPROLITES*, SUGGESTING THAT HYAENAS PERIODICALLY LIVED IN THIS PART OF THE CAVE SPACE.

... AND HERE IS A PIT-LIKE FEATURE FOUND IN SQUARE M16 DURING OUR EXCAVATIONS -- IT IS RIGHT NEXT TO WHERE THE SKELETON WAS FOUND IN 1961.

THIS IS ACTUALLY A REMNANT OF A KARSTIC CHANNEL OR TUBE, ALTHOUGH FROM THIS VIEW IT LOOKS LIKE A PIT SIMILAR TO THE ONE THAT THE CHILD WAS FOUND IN. THIS CHANNEL EXTENDS ACROSS THE FLOOR OF THE CAVE AND INTO THE BEDROCK AS WE CAN SEE ON THE NEXT PAGE.

ROC DE MARSAL
10 06 04
FOSSE
À LA BASE DE
LA COUPE OUEST
CARRÉ M16

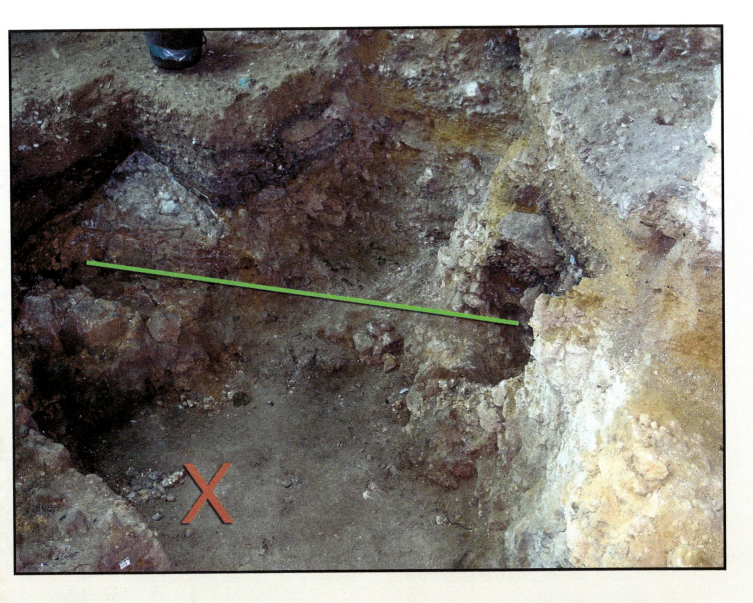

HERE'S A VIEW OF THE CENTRE OF THE CAVE FLOOR.

THE GREEN LINE INDICATES THE REMAINS OF THE ANCIENT KARSTIC CHANNEL THAT LED TO THE PIT-LIKE FEATURE WE FOUND IN SQUARE M16 (AT THE RIGHT END OF THE LINE).

THE LOCATION OF THE SKELETON IS MARKED WITH AN X.

CLOSE-UP VIEW OF THE PIT-LIKE FEATURE IN SQUARE M16 SIMILAR TO THE ONE THE CHILD WAS FOUND IN

THAT IS A VERY GOOD QUESTION.

THE BEST WAY TO ANSWER IT IS TO SHOW YOU HOW THE POSITION OF THE SKELETON RELATED TO THE SEDIMENTS IN THE PIT.

WHEN WE EXAMINED THE PIT IN SQUARE M16, IT BECAME OBVIOUS THAT THIS IS THE REMNANT OF A NATURAL KARSTIC CHANNEL, JUST LIKE THE ONE THE SKELETON WAS FOUND IN. AS YOU CAN SEE IN THIS PHOTO AND SKETCH, THIS PIT WAS FILLED WITH SEVERAL NATURAL LAYERS OF SEDIMENTS.

ACCORDING TO THE REPORTS BY LAFILLE AND BORDES, THE INFANT'S SKULL WAS LOCATED IN TWO DIFFERENT LAYERS. IN ADDITION, THE BODY WAS LAYING ON TOP OF SEDIMENTS WHICH HAD ALREADY BEGUN TO ACCUMULATE IN THE PIT.

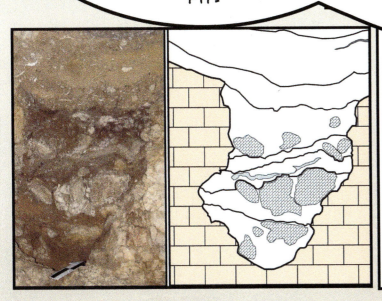

51

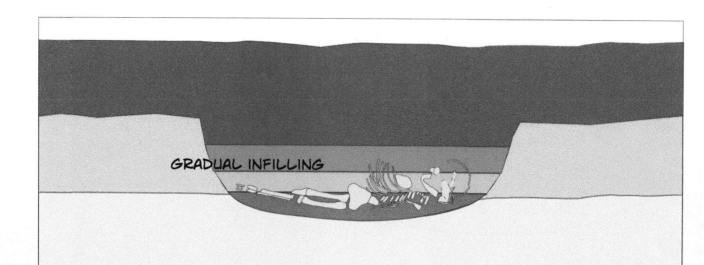

GRADUAL INFILLING

FROM LAFILLE AND BORDES' DESCRIPTIONS, IT SOUNDS VERY MUCH LIKE THE INFANT WAS COVERED BY MULTIPLE LAYERS — SIMILAR TO THIS DRAWING. THIS INDICATES SLOW, NATURAL DEPOSITIONAL PROCESSES — NOT RAPID BURIAL AT THE TIME OF DEATH, AS WOULD BE THE CASE WITH INTENTIONAL BURIAL.

SINGLE INFILLING EVENT

THE SKELETON AND SEDIMENTS OF AN INTENTIONAL BURIAL SHOULD LOOK SOMETHING LIKE THIS. HERE IT IS CLEAR THAT THE BODY WAS LAID IN THE BOTTOM OF THE PIT AND THEN COVERED OVER IMMEDIATELY BY A SINGLE HOMOGENEOUS LAYER OF SEDIMENTS.

THERE'S NO INDICATION FROM THE MICROMORPHOLOGICAL SAMPLES* THAT THIS PART OF THE CAVE WOULD BE SUBJECT TO ANY PROCESSES THAT COULD LEAD TO RAPID ACCUMULATION OF NATURAL LAYERS. IT REALLY LOOKS LIKE THE SKELETON WAS COVERED GRADUALLY OVER TIME.

NOT ONLY THAT, BUT THE SKELETON WAS MISSING ITS FEET WHEN IT WAS FOUND! THIS PROBABLY MEANS THAT THEY WERE EXPOSED FOR A PERIOD OF TIME AND WERE DAMAGED OR SCAVENGED BY A CARNIVORE.

VERA ALDEIAS--SPECIALIST IN THE ANALYSIS OF SEDIMENTS IN ARCHAEOLOGICAL SITES.

WOW, THAT REALLY DOESN'T SOUND LIKE TYPICAL BURIAL BEHAVIOR, HUH?

BUT, WAS THE BODY OF THE INFANT ARRANGED IN A SPECIAL POSITIONED FOR BURIAL?

THE ROC DE MARSAL CHILD IS ONE OF THE MOST COMPLETE NEANDERTHAL SKELETONS EVER DISCOVERED.

THE PUBLISHED DESCRIPTIONS, PHOTOS, AND DRAWINGS ALL SHOW THAT THE CHILD WOULD HAVE BEEN LYING FACE-DOWN WITH THE LOWER LEGS FLEXED BACKWARDS AND TO THE RIGHT.

...SO THE BODY WOULD HAVE LOOKED SOMETHING LIKE THIS:

THIS IS VERY ATYPICAL POSITIONING FOR A DELIBERATE BURIAL.

IN DELIBERATE BURIALS THE BODY IS OFTEN ARRANGED IN A SPECIAL POSITION.

MANY ARE CAREFULLY PLACED IN EITHER THE *FETAL POSITION* OR THE *SUPINE POSITION*:

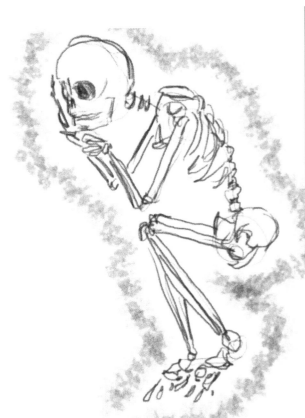

"FETAL POSITION": CURLED UP WITH KNEES TO CHEST.

IT SHOULD BE NOTED, HOWEVER, THAT WHILE SOME PEOPLE ARE INTENTIONALLY BURIED IN THIS POSITION IT IS ALSO THE POSITION A PERSON ASSUMES WHEN THEY ARE DYING OF EXPOSURE.

"SUPINE POSITION": STRETCHED OUT AND LYING ON THE BACK.

NO NEANDERTHAL SKELETONS HAVE BEEN FOUND IN THIS POSITION.

OK. LET'S RECAP. HERE'S WHAT THE EVIDENCE SAYS ABOUT THE ROC DE MARSAL INFANT:

1. THE PIT WHERE THE SKELETON WAS FOUND IS PART OF A NATURAL DEPRESSION WITHIN THE BEDROCK. IT WASN'T PURPOSELY DUG FOR A BURIAL.

2. THE STRATIGRAPHY SHOWS THAT THE SKELETON WAS COVERED BY AT LEAST TWO SEPARATE LAYERS OF SEDIMENT. THIS MEANS THAT THE INFILLING IN THE PIT WAS THE RESULT OF NATURAL DEPOSITION ... IT WASN'T DONE ALL AT ONCE, AS YOU'D EXPECT WITH A BURIAL.

3. THE POSITION OF THE SKELETON IN THE DEPRESSION AND THE DAMAGED LOWER EXTREMITIES ARE NOT AT ALL CONSISTENT WITH BURIAL BEHAVIOR.

4. THERE **WERE** ARTIFACTS FOUND IN THE SEDIMENTS WITH THE SKELETON, BUT THE SAME TYPES AND QUANTITIES OF ARTIFACTS ARE FOUND THROUGHOUT ALL THE LAYERS, SO THERE IS NO EVIDENCE OF "GRAVE GOODS."

SO IT LOOKS LIKE THE ROC DE MARSAL CHILD IS AN EXCELLENT EXAMPLE OF A NEARLY COMPLETE NEANDERTHAL CHILD, BUT THERE IS NO EVIDENCE THAT IT WAS DELIBERATE BURIED.

WHILE THIS DOESN'T PROVE THAT NEANDERTHALS NEVER BURIED THEIR DEAD, IT CERTAINLY WEAKENS ANY ARGUMENT THAT THEY DID, AND AT LEAST AT ROC DE MARSAL THERE IS NO EVIDENCE THAT NEANDERTHALS ENGAGED IN ANY KIND OF RITUAL BEHAVIOR.

THERE ARE MANY OTHER FASCINATING QUESTIONS THAT SITES LIKE ROC DE MARSAL CAN HELP US ANSWER. FOR EXAMPLE:

- DID NEANDERTHALS USE FIRE AND DID THEY KNOW HOW TO MAKE IT?

- WHAT WAS THEIR STONE TOOL TECHNOLOGY LIKE?

- HOW DID THEY HUNT AND BUTCHER ANIMALS?

- AND, HOW DO ARCHEOLOGISTS EXCAVATE A SITE?

BUT, WE WILL HAVE TO ADDRESS THESE QUESTIONS IN OTHER BOOKS.

GLOSSARY OF IMPORTANT TERMS

ANTHROPOGENIC: 'ANTHROPOGENIC' MEANS CAUSED OR CREATED BY HUMANS -- AS OPPOSED TO NATURAL PROCESSES. SOME STRATA AT AN ARCHAEOLOGICAL SITE WILL BE COMPOSED OF MAINLY NATURAL SEDIMENTS (SILT DEPOSITED BY THE WIND OR ROCKS THAT FALL FROM A CAVE ROOF OVER TIME), BUT SOME STRATA WILL INCLUDE SIGNIFICANT ANTHROPOGENIC COMPONENTS LIKE THE ASH AND CHARCOAL FROM PEOPLE'S CAMPFIRES.

CONTEXT: IN ARCHAEOLOGY, THE TERM 'CONTEXT' REFERS TO THE PHYSICAL CIRCUMSTANCES IN WHICH AN ARTIFACT OR FEATURE ARE SITUATED IN AN ARCHAEOLOGICAL SITE: ITS PHYSICAL RELATIONSHIP TO EVERYTHING AROUND IT. THIS INCLUDES A WIDE RANGE OF THINGS. FOR EXAMPLE: ITS 3-DIMENSIONAL POSITION IN SPACE, THE LAYER IT WAS BURIED IN, OR WHAT OTHER ARTIFACTS OR FEATURES WERE ASSOCIATED WITH IT.

COPROLITE: THIS IS FOSSILIZED FECES OR DUNG. THIS INCLUDES HUMAN FECES WHICH CAN, ON RARE OCCASIONS, BE PRESERVED IN ARCHAEOLOGICAL DEPOSITS. HUMAN COPROLITES ARE TYPICALLY GREAT SOURCES OF INFORMATION ABOUT PREHISTORIC DIET.

CORE: THIS IS A TERM ARCHAEOLOGISTS USE TO REFER TO NODULES OF STONE THAT PEOPLE REMOVED FLAKES FROM IN ORDER TO USE THOSE FLAKES AS TOOLS. AT SITES LIKE ROC DE MARSAL, WE FIND DOZENS OF FLINT CORES THAT HAD MANY FLAKES REMOVED FROM THEM. HOWEVER, THE STONE TOOL COLLECTIONS FROM NEANDERTAL SITES LIKE THIS ARE COMPOSED MAINLY OF THE FLAKES THAT WERE REMOVED FROM THE CORES: SOME OF WHICH HAD BEEN TURNED INTO RETOUCHED TOOLS.

GEOARCHAEOLOGY: USING GEOSCIENCE APPROACHES (GEOLOGY, GEOPGYSICS, PEDOLOGY) TO SOLVE ARCHAEOLOGICAL PROBLEMS. COMMONLY, HOWEVER, IT USES ARCHAEOLOGICAL DATA TO COMPREHEND GEOLOGICAL ISSUES, SUCH AS EARTHQUAKE AND FLOOD FREQUENCIES.

GEOMORPHOLOGY: THIS IS THE BRANCH OF THE GEOSCIENCES THAT DEALS WITH THE ORIGIN AND DEVELOPMENT OF LANDFORMS AND THE PROCESSES THAT PRODUCE THEM.

KARST: THESE ARE THE LANDFORMS ASSOCIATED WITH THE DISSOLUTION OF TYPICALLY CARBONATE ROCKS LIKE LIMESTONE AND DOLOMITE. CLASSIC LANDFORMS ARE CAVES AND SINKHOLES, AS WELL AS DISSOLUTION FEATURES THAT DEVELOP ON THE SURFACE OF CARBONATE ROCKS.

MICROMORPHOLOGY: THIS IS STUDY OF SOILS AND ARCHAEOLOGICAL MATERIALS USING PETROGRAPHIC THIN SECTIONS MADE FROM SMALL, UNDISTURBED SAMPLES OF THE SEDIMENTS TAKEN FROM ARCHAEOLOGICAL SITES.

PALEOLITHIC: THIS REFERS TO VERY ANCIENT TIMES. IN EUROPE THE PALEOLITHIC PERIOD BEGINS WHEN PEOPLE FIRST ARRIVE HERE (AROUND 1.4 MILLION YEARS AGO) AND ENDS WHEN PEOPLE BEGIN FARMING (AROUND 7500 YEARS AGO). THE PALEOLITHIC IS DIVIDED UP INTO THE LOWER PALEOLITHIC (ASSOCIATED WITH HOMO ERECTUS), THE MIDDLE PALEOLITHIC (ASSOCIATED WITH NEANDERTHALS), AND THE UPPER PALEOLITHIC (WHEN MODERN HUMANS ARRIVED).

STRATIGRAPHY: STRATIGRAPHY REFERS TO THE NATURAL LAYERING OF SEDIMENTS THAT OCCURS WHEN THEY ACCUMULATE AT A LOCATION OVER LONG PERIODS OF TIME. WE OFTEN USE THE WORD 'STRATUM" TO REFER TO A LAYER OR THE PLURAL, 'STRATA', TO REFER TO MULTIPLE LAYERS. DIFFERENT LAYERS OR 'STRATA' IN THE STRATIGRAPHY OF AN ARCHAEOLOGICAL SITE WILL HAVE DIFFERENT CHARACTERISTICS, LIKE COLOUR AND TEXTURE, BECAUSE THE WAY THEY WERE DEPOSITED CHANGED OVER TIME: SOME WERE DEPOSITED BY WIND, SOME BY FLOWING WATER, AND SOME BY GRAVITY.

ZOOARCHAEOLOGY: THIS IS THE STUDY OF ANIMAL REMAINS RECOVERED FROM ARCHAEOLOGICAL SITES AND IS ALSO CALLED 'FAUNAL ANALYSIS'. ZOOARCHAEOLOGICAL STUDIES CAN TELL US A LOT ABOUT WHAT PEOPLE WERE EATING IN PREHISTORY. THEY CAN ALSO TELL US A LOT ABOUT WHAT THE ENVIRONMENT WAS LIKE IN A REGION AT DIFFERENT TIMES BECAUSE OF WHAT SPECIES OF ANIMALS WERE AROUND AT THESE TIMES.

FURTHER READING

BAR-YOSEF, OFER & BERNARD VANDERMEERSCH, ET AL, "THE EXCAVATIONS IN KEBARA CAVE, MOUNT CARMEL", CURRENT ANTHROPOLOGY 33.5 (1992), PP 497-546

BELFER-COHEN, ANNA AND ERELLA HOVERS. 1992. IN THE EYE OF THE BEHOLDER: MOUSTERIAN AND NATUFIAN BURIALS IN THE LEVANT. CURRENT ANTHROPOLOGY VOL. 33, NO. 4: 463-71.

BINANT, PASCALE. 1991. LA PRÉHISTOIRE DE LA MORT : LES PREMIÈRES SÉPULTURES EN EUROPE. ÉDITIONS ERRANCE, PARIS.

CHASE, PHILLIP & HAROLD L. DIBBLE, 1987. MIDDLE PALEOLITHIC SYMBOLISM: A REVIEW OF CURRENT EVIDENCE AND INTERPRETATIONS. JOURNAL OF ANTHROPOLOGICAL ARCHAEOLOGY 6, 263-296.

DEFLEUR, ALBAN. 1993. LES SÉPULTURES MOUSTÉRIENNES. C.N.R.S. ÉDITIONS, PARIS.

DIBBLE, H. L., V. ALDEIAS, P. GOLDBERG, S. P. MCPHERRON, D. SANDGATHE, & T. E. STEELE. 2014. A CRITICAL LOOK AT EVIDENCE FROM LA CHAPELLE-AUX-SAINTS SUPPORTING AN INTENTIONAL NEANDERTAL BURIAL. JOURNAL OF ARCHAEOLOGICAL SCIENCE.

GARGETT, ROBERT H., 1989. GRAVE SHORTCOMINGS: THE EVIDENCE FOR NEANDERTAL BURIAL. CURRENT ANTHROPOLOGY 30, 157-190.

GARGETT, ROBERT H., 1999. MIDDLE PALAEOLITHIC BURIAL IS NOT A DEAD ISSUE: THE VIEW FROM QAFZEH, SAINT-CÉSAIRE, KEBARA, AMUD, AND DEDERIYEH. JOURNAL OF HUMAN EVOLUTION 37, 27-90.

GOLDBERG, P., V. ALDEIAS, H. L. DIBBLE, S. P. MCPHERRON, D. SANDGATHE & A. TURQ, 2013. TESTING THE ROC DE MARSAL NEANDERTAL "BURIAL" WITH GEOARCHAEOLOGY. ARCHAEOLOGICAL AND ANTHROPOLOGICAL SCIENCES, 1-11.

HARROLD, FRANCIS B. 1980. A COMPARATIVE ANALYSIS OF EURASIAN PALEOLITHIC BURIALS. WORLD ARCHAEOLOGY 12(2)195-211.

HENSHILWOOD, CHRISTOPHER & CURTIS MAREAN, 2006. REMODELLING THE ORIGINS OF MODERN HUMAN BEHAVIOUR. THE PREHISTORY OF AFRICA: TRACING THE LINEAGE OF MODERN MAN, 31-46.

LAVILLE, H., RIGAUD, J.-P. & SACKETT, J. (1980) ROCK SHELTERS OF THE PERIGORD: GEOLOGICAL STRATIGRAPHY AND ARCHAEOLOGICAL SUCCESSION, NEW YORK: ACADEMIC PRESS.

LINDLY, J.M. & G.A. CLARK. 1990. SYMBOLISM AND MODERN HUMAN ORIGINS. CURRENT ANTHROPOLOGY VOL. 31, NO. 3:233-61.

MCBREARTY, S. & BROOKS, A.S., 2000. THE REVOLUTION THAT WASN'T: A NEW INTERPRETATION OF THE ORIGIN OF MODERN HUMAN BEHAVIOR. JOURNAL OF HUMAN EVOLUTION 39, 453-563.

RENDU, W., C. BEAUVAL, I. CREVECOEUR, P. BAYLE, A. BALZEAU, T. BISMUTH, L. BOURGUIGNON, G. DELFOUR, J.-P. FAIVRE, F. LACRAMPE-CUYAUBÈRE, C.TAVORMINA, D. TODISCO, A. TURQ, & B. MAUREILLE 2014 EVIDENCE SUPPORTING AN INTENTIONAL NEANDERTAL BURIAL AT LA CHAPELLE-AUX-SAINTS. PROCEEDINGS OF THE NATIONAL ACADEMY SCIENCES 111, 81-86.

SANDGATHE, D.M., H.L. DIBBLE, P. GOLDBERG, S.P. MCPHERRON, 2011. THE ROC DE MARSAL NEANDERTAL CHILD: A REASSESSMENT OF ITS STATUS AS A DELIBERATE BURIAL. JOURNAL OF HUMAN EVOLUTION 61, 243-253.

TURQ, ALAIN. 1989. "LE SQUELETTE DE L'ENFANT NÉANDERTALIEN DU ROC DE MARSAL : LES DONNÉES DE FOUILLES" PALÉO 1:47-54.

VANDERMEERSCH, BERNARD. 1976. LES SÉPULTURES NÉANDERTALIENNES. IN DE LUMLEY, HENRY, EDITOR, LA PRÉHISTOIRE FRANÇAISE, VOLUME 1. PARIS: CENTRE NATIONALE DE LA RECHERCHE SCIENTIFIQUE, PP. 725-7.

WHITE, W. B. & CULVER, D. C., EDS. (2012) ENCYCLOPEDIA OF CAVES, 2ND EDITION, 2 ED., SAN DIEGO: ACADEMIC PRESS.